Earth Science Week 2016: Our Shared Geoheritage

AF879589

Highlights Report

ISBN: 978-0-913312-56-8

American Geosciences Institute
4220 King Street
Alexandria, VA 22302 U.S.A.
www.americangeosciences.org
703-379-2480

If you have comments concerning this report, please contact:
Ed Robeck, Ph.D.
Director, Department of Education and Outreach
Director, Center for Geoscience and Society
American Geosciences Institute
703-379-2480 x245
ecrobeck@americangeosciences.org

News Clips www.earthsciweek.org/highlights

Because of the large and increasing number of news clippings citing Earth Science Week activities and resources, the print edition of the report no longer includes clippings. To view the hundreds of press releases and news items promoting awareness of Earth Science Week each year, please go online to www.earthsciweek.org/highlights. Thank you for helping us in our efforts to conserve resources and protect the environment.

Table of Contents

Front cover: 2016 ESW poster design for AGI by Angela Terry. Front and back cover background ©Shutterstock.com/Pakhnyushcha.

EARTH SCIENCE WEEK

AGI american geosciences institute
connecting earth, science, and people

Highlights Report: Earth Science Week 2016

Finalist entry to the Earth Science Week 2016 Photography Contest by Katy Sprinkle.

Introduction

Held October 9-15, 2016, the 19th annual **Earth Science Week** celebrated the theme of **"Our Shared Geoheritage."** The 2016 event, like past celebrations, promoted public and professional awareness of Earth science in education and society.

"Earth Science Week 2016 celebrates 'Our Shared Geoheritage,' a concept that has become a rallying cry for geoscience-related efforts worldwide," says Geoff Camphire, AGI's Manager of Outreach. "Geoheritage awareness is essential for all citizens of the planet, who are the 21st-century inheritors, managers, and protectors of a vast wealth of irreplaceable natural treasures, billions of years in the making." Earth Science Week 2016 learning resources and activities promoted awareness of geoheritage.

AGI organizes Earth Science Week as a service to member societies, with generous help from partners that provide funding, donate materials, organize events, and publicize the event. Funding partners in 2016 included the U.S. Geological Survey (USGS); American Association of Petroleum Geologists (AAPG) Foundation; National Park Service; National Aeronautics and Space Administration (NASA); American Geophysical Union (AGU); Geological Society of America (GSA); ExxonMobil; Association of American State Geologists (AASG); Society for Mining, Metallurgy and Exploration (SME); AmericaView; Archaeological Institute of America (AIA); Howard Hughes Medical Institute; and Esri.

Earth Science Week **participation remained strong.** People in all 50 states and more than 18 countries participated in events and activities. The Earth Science Week website received over 675,000 page views in 2016. In addition, hundreds of people nationwide actively participated in the program's visual arts, essay, and photography contests.

Earth Science Week events ranged from educators teaching Earth science activities in their classrooms to open houses held at major USGS field stations. A detailed list of events can be found in the second half of this report. This list represents only events reported directly to AGI, so please notify Earth Science Week staff if your participation is not listed.

Additional events are highlighted on the Earth Science Week website (www.earthsciweek.org/highlights), which features press releases and other items by members of the geoscience community, as well as news media promoting Earth Science Week. Television and radio news programs covered Earth Science Week on local stations in several states. Each year, web, print, and broadcast media coverage of Earth Science Week reaches more than **50 million people.**

This report is designed to give an overview of the activities organized by AGI and other groups for Earth Science Week. We hope this information on 2016 events and publicity inspires you to develop your own activities next year. Please visit www.earthsciweek.org for event planning, materials, resources, and support. Contact Earth Science Week staff at info@earthsciweek.org for assistance in planning for Earth Science Week.

Summary of Activities

Winning entry to the Earth Science Week 2016 Photography Contest by Lily Wolfe.

Overview

More than 50 million people gained a new awareness of the geosciences through the 19th annual Earth Science Week. People worldwide learned about Earth science through program promotions, education efforts, news media, and Internet coverage. The event celebrated the theme "Our Shared Geoheritage" by promoting awareness of the many ways that science helps us understand, appreciate, and make the most of our geoscience heritage, or, as it is commonly known worldwide, "geoheritage."

Earth Science Week events ranged from individual teachers and classes conducting playground science projects to activities worldwide. Many activities were organized around the days of Earth Science Week that routinely are set aside times for areas of special focus:

- Geocachers worldwide used GPS devices to find and learn about sites of geoscientific significance on **International EarthCache Day,** October 9.
- AGI's "Big Ideas of Earth Science" videos and activities allowed students to explore core concepts on **Earth Science Literacy Day,** October 10.
- Schools nationwide held activities for **Earth Observation Day** and **No Child Left Inside Day** on October 11, encouraging students to explore remote sensing and other geoscience topics.
- The seventh annual **National Fossil Day,** October 12, invited young people to learn about paleontology at parks and other locations across the country.
- **Geoscience for Everyone Day,** October 13, reached out to underrepresented groups, including women, minorities, and people with a range of abilities.

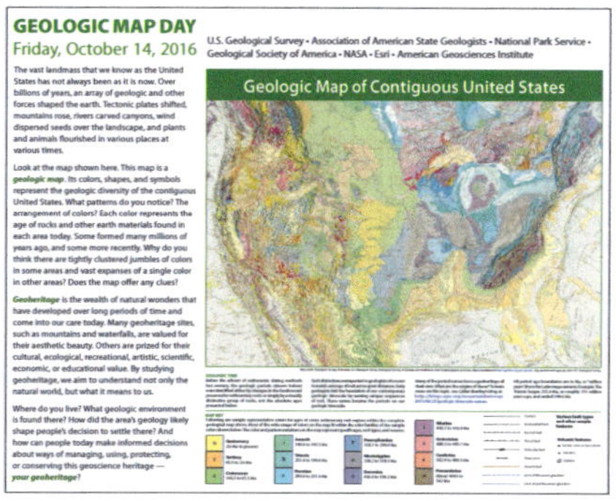

Image credit: AGI

- The fifth annual **Geologic Map Day,** October 14, offered resources to help young people explore the study, uses, and importance of geologic maps.
- Finally, program participants once again explored archaeological wonders on **International Archaeology Day,** October 15.

Internationally, students and others in Australia, Brazil, Cameroon, Canada, Colombia, the Czech Republic, Guyana, Ireland, Jamaica, Mexico, Peru, Russia, Saint Lucia, Singapore, Trinidad & Tobago, the United Arab Emirates, the United Kingdom, and Uruguay participated in events, explorations, and activities during and around Earth Science Week. Hundreds of essayists, artists, and shutterbugs once again entered the annual Earth Science Week contests.

Key Partnerships and Efforts

AGI Education Department staffer, Celia Thomas, showcases an ESW toolkit.
Image credit: Juliet Crowell/AGI

Earth Science Week's success depends on the collaboration of key partners. In 2016, AGI forged partnerships with several organizations and pursued signature initiatives:

AGI recently expanded eligibility for its Earth Science Week photo contest to allow members of **AGI International Affiliates** and international members of AGI Member Societies to participate. Previously open only to U.S. residents, the photo contest has always been a major part of Earth Science Week.

Educators seeking teaching resources and other support were directed by AGI to the **American Association of Petroleum Geologists** (AAPG) and the AAPG Foundation, both longtime supporters of Earth Science Week. In addition, Earth Science Week promoted awareness of AAPG's Distinguished Lecturer program. AAPG Student Chapters received kits. Earth Science Week's 2016 activity calendar featured a classroom activity by AAPG. Program participants were encouraged to read and use "Visiting Geoscientists: An Outreach Guide for Geoscience Professionals," a handbook co-produced by AGI and AAPG's Youth Education Activities Committee. Program participants were encouraged to compete to win the AAPG Teacher of the Year award for a U.S. K-12 geoscience educator. AAPG's "Natural Gas Formation" activity was featured in the Earth Science Week 2016 activity calendar.

AGI supported the **Austin Earth Science Week Consortium**'s celebration in Texas in 2016 by donating posters and other educational materials. Hundreds of middle-school students attended a career fair and participated in activities.

The **American Geophysical Union** (AGU) continued its role as a supporting program partner in 2016 with the contribution of funds as well as expertise. Earth Science Week's 2016 activity calendar featured AGU's "Watch Out for Landslides" classroom activity. AGU's annual meetings, professional development workshops, programs for college students, print and electronic resources, GIFT workshops, and "LEARN With AGU" video series were promoted through the Earth Science Week e-newsletter, website, and activity calendar. Earth Science Week staff also presented at AGU's Fall Meeting in San Francisco. In addition, AGU provided an "Exploring Earth and Space" coloring book for the educator kit.

Earth Science Week 2016 promoted awareness of the **American Institute of Professional Geologists** (AIPG), an AGI member society that advocates for geologists and certifies their credentials. AIPG offers several PowerPoint presentations online for free download, presenting career information for young, newly graduated geoscientists. AIPG also provided a geologic-timescale bookmark for the educator kit.

Program participants were encouraged to go online to conduct a new activity called "Climate at a Glance: From Local to National Scale" that the **American Meteorological Society** (AMS) created in cooperation with NOAA's National Centers for Environmental Information. The activity introduces students to the NOAA Climate at a Glance website, which allows real-time analysis of monthly temperature and precipitation data nationwide. AMS also provided a flyer detailing this activity for the Earth Science Week 2016 Toolkit.

Earth Science Week 2016 directed participants' attention to the annual celebration of Earth Observation Day, a STEM educational outreach event of **AmericaView** and its partners, which engage students and teachers in remote sensing as an exciting and powerful educational tool. AmericaView also provided a "Memory Game" poster dealing with remote sensing and geoheritage for the educator kit.

The **Archaeological Institute of America** (AIA), a continuing Earth Science Week partner and supporter, provided a classroom activity on "Tree Rings and Ancient

Climatic Conditions" for the program's 2016 activity calendar. In addition, the program promoted awareness of and participation in AIA's International Archaeology Day, which takes place annually on the final day of Earth Science Week.

Competitors were recruited to submit artworks or videos to celebrate **Arctic Science Summit Week** in 2016. The event's contest encouraged geoscience learning among K-12 students preparing for Earth Science Week.

Earth Science Week directed participants' attention to the **Association of American Geographers** (AAG), an AGI member society that offers an array of web resources for K-12 and college-level instruction. Materials support geographic approaches to Earth science education. For example, Geographic Advantage, an educational companion for the National Research Council's "Understanding the Changing Planet," outlines teaching strategies and investigations that show students how geographers use maps to explore environmental change.

The **Association of American State Geologists** partnered with AGI and the USGS to support Geologic Map Day during Earth Science Week 2016. State geologists nationwide made geologic maps of their states available on their websites for students to use in classroom activities on Geologic Map Day.

Encouraging educators to make use of the resources offered by the **Association of Environmental and Engineering Geologists** (AEG) to its members, Earth Science Week 2016 turned a spotlight on AEG's technical publications, section and chapter meetings, and special educator sessions at the AEG annual meeting.

Earth Science Week promoted awareness of a website of great value to educators, AGI's **Center for Geoscience & Society.** The center enhances geoscience awareness across all sectors of society by generating new approaches to building geoscience knowledge, engaging the widest possible range of stakeholders, and creatively promoting existing and new resources and programs.

Program participants learned about three online videos by the **Center for Ocean Sciences Education Excellence** (COSEE) depicting dramatic changes in Alaska's marine ecosystems through interviews with scientists. The videos were produced by COSEE Alaska in cooperation with other geoscience organizations.

The **Critical Zone Observatories** provided an informational flyer on this National Science Foundation program, including a classroom activity on the Valles Caldera, for the Earth Science Week Toolkit in 2016.

Finalist entry to the Earth Science Week 2016 Visual Arts Contest by David De Costa.
Image credit: David De Costa

Earth Science Week participants were encouraged to celebrate **Earth Day** in April 2016 with classroom activities, experiments, and investigations exploring how the world works. Because Earth Science Week offers education materials, information, and tools throughout the year, school audiences were urged to make use of tools highlighting the theme of "Our Shared Geoheritage."

Leading the charge to incorporate GIS (geographic information system) technology and mapping software in Earth science education, **Esri** provided educators with useful resources and information during Earth Science Week 2016. For example, Esri launched an online educational resource called GeoInquiries for Earth Science. These free, bite-sized activities make use of topical web maps and lessons tied to standards and key textbooks. Earth Science Week, in turn, promoted GIS Day, November 16, to raise awareness of geographic information systems. GIS Day 2016 events allowed users of geographic information systems across the country and around the world to demonstrate real-world applications that make a difference in society. Esri, a GIS Day sponsor and longtime Earth Science Week partner, also served as a partner in the Geologic Map Day initiative in 2016. In addition, Earth Science Week's 2016 activity calendar featured a "Cracked Plates & Tectonics" activity from Esri, and the educator kit contained an Esri flyer detailing an activity on "Topography and Our National Heritage."

ExxonMobil, a longtime Earth Science Week partner, continued its support of the program. ExxonMobil was a major supporter of and participant in the program, including Earth Science Week Houston 2016.

Through Earth Science Week, participants learned about educational resources and programs of the **Geological Society of America** (GSA), a longtime program partner. Featured education and outreach programs included the Teacher Advocate Program, Teacher GeoVenture trips, the Distinguished Earth Science Teacher in Residence program, and GSA's GeoCorps America program. GSA also organized International EarthCache Day at the start of Earth Science Week 2016, provided a learning activity for the calendar, and contributed as an active partner in the Geologic Map Day initiative. Earth Science Week's 2016 activity calendar included GSA's "Exploring Geoheritage Through EarthCaching" activity. The educator kit also featured a GSA bookmark.

The **Geological Society of London** (GSL), an AGI member society, provided two online resources for learning about key geoscience topics. Electronic map-based resources were the focus of GSL's Plate Tectonics page. In addition, a site was launched to accompany GSL's Rock Cycle online module.

Program participants were directed to **Geology.com,** a major Earth Science Week partner which provides a variety of geoscience materials including daily Earth science news, maps, an online dictionary of Earth science terms, and information on geoscience careers, as well as resources for teachers, including links to lesson plans from major Earth science organizations. Geology.com, in turn, covered Earth Science Week announcements, programs, and activities throughout the year.

Google, an ongoing Earth Science Week partner, contributed a "Geoheritage Via Google Street View" activity to the Earth Science Week 2016 activity calendar.

The **Howard Hughes Medical Institute** (HHMI), a continuing program partner, provided key materials for the Earth Science Week 2016 Toolkit. Prized components of this year's kit were a two-disk DVD collection of HHMI lectures on "Patterns and Processes in Ecology" and a related poster, plus a wildlife restoration postcard.

To help teachers and students delve into the science behind current events, Earth Science Week directed them to "Recent Earthquake Teachable Moments," a website of the **Incorporated Research Institutions for Seismology** (IRIS). These resources — including PowerPoint presentations, animations, and visualizations, as well as links to Spanish-language materials and USGS data — dealt with, for example, the magnitude-7.2 earthquake that shook Kamchatka, Russia, on January 30, 2016.

Earth Science Week 2016 promoted awareness of a new free app by the **Lamont-Doherty Earth Observatory,** called "Polar Explorer: Sea Level," which lets users explore a series of maps of the planet, from the deepest trenches in the oceans to the ice at the poles. Users see how ice, the oceans, precipitation, and temperatures have changed over time and hear scientists explain.

The entrance to the Smithsonian Museum of Natural History publicizing National Fossil Day.
Image credit: Celia Thomas/AGI

The **Minerals Education Coalition** (MEC) of the **Society for Mining, Metallurgy, and Exploration,** an AGI member society, supported Earth Science Week in 2016. MEC provided a flyer featuring the educational "MEC Mining & Money Catcher" game for the 2016 educator kit. Earth Science Week's 2016 activity calendar featured a "Density of Minerals" classroom activity from MEC. And program participants received information about a series of MEC podcasts with industry experts, showing students how an interest in STEM subjects can lead to a rewarding career in the mining industry.

Earth Science Week directed participants' attention to a joint effort by AGI and the **National Association of Geoscience Teachers** (NAGT) to strengthen implementation of the Next Generation Science Standards (NGSS) at the state level. Following an NGSS Summit of education leaders in April 2015, science educators were invited to participate in a free webinar and join the discussion online. Supporting implementation of the NGSS in Earth and Space Sciences requires partnerships between school districts, state education departments, and two- and four-year colleges and universities. A consortium of Washington State organizations currently is working to improve science learning for all students statewide by creating a new vision for STEM teacher preparation aligned with NGSS. This webinar focused on building the consortium and progress so far. In addition, program participants learned about NAGT's Outstanding Earth Science Teacher Awards, Dorothy Stout Professional Development Grants, geologic mapping lessons online, and other offerings.

NASA, a founding partner of Earth Science Week, provided a "National Parks from Space" DVD, offering virtual visits to park sites from the vantage point of space, along with additional information for students and teachers, for the Earth Science Week 2016 Toolkit. Also included in the kit were an illustrated 32-page "Tour of the Electromagnetic Spectrum" booklet, a GLOBE mini-flyer on satellite observations of clouds, and a flyer on Earth's energy budget. Earth Science Week's 2016 activity calendar also featured an "Exploring Geoheritage From Space" classroom activity from NASA. Earth Science Week promoted awareness of NASA's online offerings, such as SciJinks and Climate Kids, throughout the year. Also, Earth Science Week participants took part in the first-ever Earth Observation Day, a STEM outreach event celebrating the Landsat mission of USGS and NASA.

The **National Earth Science Teachers Association** (NESTA), a longtime Earth Science Week partner, continued its vital role in helping AGI promote excellence in geoscience education. At the National Science Teachers Association Annual Conference, the NESTA Reception included a ceremony during which a teacher was given the Edward C. Roy, Jr. Award for Excellence in K-8 Earth Science Teaching. In addition, NESTA members received copies of the Earth Science Week 2016 poster in their association newsletter.

Earth Science Week raised awareness of **National Environmental Education Week** (EE Week), the nation's largest environmental education event. Focusing on "Greening STEM," EE Week connected educators with resources to promote K-12 students' understanding of the environment.

Earth Science Week promoted the **National Ground Water Association**'s (NGWA) Ground Water Awareness Week in March 2016 as well as NGWA's Protect Your Groundwater Day program in September 2016, advocating water conservation and safety. The AGI member society offers Ground Water Adventures, a website providing educational activities for young people.

Earth Science Week's 2016 activity calendar featured a "What-a-Cycle" activity from the National Oceanic and Atmospheric Administration (NOAA). In addition, Earth Science Week 2016 directed participants to NOAA's online multimedia education resources, including lesson plans, real-world data, instructional games, videos, and more.

A longtime Earth Science Week partner, the **National Park Service** (NPS) continued for the seventh year a major component to its involvement in Earth Science Week. National Fossil Day, established as a celebration to take place annually on the Wednesday of Earth Science Week, once again reached millions of people. Earth Science Week promoted awareness of NPS's National Parks BioBlitz, during which teams of scientists, naturalists, students, and volunteers documented living creatures of national parks. Similarly, NPS hosted a PaeloBlitz event for which program participants were recruited. Program participants also learned about the NPS National Natural Landmarks program, which recognizes and encourages the conservation of sites that contain outstanding biological and geological resources. NPS videos on climate change were made available to program participants. NPS's centennial celebration was a major focus of promotions. Posters illuminating both geologic and atmospheric resources of the parks as well as centennial information, appearing in the Earth Science Week 2016 Toolkit, successfully continued the series of park posters produced collaboratively by the agency and AGI. The agency also served as a partner in the Geologic Map Day initiative in 2016. Earth Science Week's 2016 activity calendar also featured a classroom activity on "Rock Art in the National Parks."

Finalist entry to the Earth Science Week 2016 Visual Arts Contest by Shrish Karthikeyan.
Image credit: Shrish Karthikeyan

Earth Science Week participants learned about soil education resources offered online by the **National Resource Conservation Service** (NRCS). Resources for the elementary level include lesson plans, links to soil education websites, and even soil songs.

For Earth science teachers and students searching for the most up-to-date information on climate change, Earth Science Week 2016 highlighted a **National Science Foundation** (NSF) website. The site features an NSF report summarizing the current state of knowledge about climate change, as well as resources dealing with related news, discoveries, statistics, and publications.

Earth Science Week partnered with the **National Science Teachers Association** (NSTA) again in 2016, reaching science educators nationwide. Program participants learned about "Freebies for Science Teachers" on the NSTA website. Also, AGI exhibited once again at the NSTA Annual National Conference, sharing Earth Science Week and other education material for science teachers.

The Nature Conservancy, which offers informational resources ideal for educators aiming to teach about a wide range of geoscience topics, was promoted through

Earth Science Week. Videos, webcasts, articles, and other materials conveyed the work of scientists engaged in conservation efforts around the world.

Earth Science Week raised awareness of Houston's **Offshore Energy Center** (OEC), which offered educational resources for Earth science teachers. OEC's Education Outreach program provided, for instance, an online Career Interest Profile that students could fill out to learn which energy careers dovetail with what they enjoy doing.

As detailed in the Earth Science Week newsletter, the **Oklahoma Energy Resources Board** (OERB) provides materials and services to improve the lives of Oklahomans and others through education and restoration. OERB educational resources such as the "Career Paths" web page provide information on careers in the Earth sciences.

Gearing up for National Fossil Day, Earth Science Week directed program partners' attention to the **Paleontological Research Institution** (PRI), an AGI member society providing education materials and opportunities for science teachers and students at all grade levels. The online Teacher Friendly Guide, for example, gives brief geologic histories of every region of the United States.

Partners in Resource Education (PRE), a longtime program partner, provides activities focusing on the geoscience of conservation. The consortium of seven federal agencies educates thousands of young people, introduces them to natural resource careers, and cultivates the next generation of land and water stewards. In 2016, PRE collaborated to promote awareness of Earth Science Week, and vice versa.

Schlumberger Excellence in Educational Development (SEED) is a nonprofit education program that empowers Schlumberger employee volunteers and educators to share their passion for learning and science with students. In addition to promoting awareness of SEED's site and other resources, AGI partnered with the program to launch a joint initiative, the SEED Earth Science Week Online Toolkit, to provide geoscience education resources in both Spanish and English in 2010. With new resources added periodically since then, visitors were able to view, download, and print over 70 lessons, posters, fact sheets and other items, each offered in both Spanish and English, until the site was discontinued in 2016.

For teachers aiming to "shake up" education, Earth Science Week shone a spotlight on the **Seismological Society of America** (SSA). SSA's website provided seismic eruption models, wave animations, plate tectonics simulations, information on tsunamis, and more. SSA also offered publications, information on seismology careers, a distinguished lecturer series, and an electronic encyclopedia of earthquakes.

Dino Twister game and Trackway for National Fossil Day activities.
Image credit: Celia Thomas/AGI

Program participants received information about **Smithsonian Education**, which offers a fascinating exploration of Earth's soil with its "Dig It! The Secrets of Soil" exhibition. Information, videos, expert instruction, and activity sheets are available online.

Earth Science Week participants were encouraged to take advantage of offerings of the **Society of Exploration Geophysicists** (SEG), which provides programs for educators and students. For example, the distinguished lecturer series and honorary lecturer series both enabled students to meet professional geophysicists, learn about groundbreaking research in the field of seismic research, and obtain valuable career information.

Advanced by the **Society of Petroleum Engineers** (SPE), the Energy4Me program offers teachers a collection of tools for teaching about oil, gas, and other energy sources, including classroom activities, experiments, and presentations, as well as teacher workshops and energy education materials for the classroom.

The **Soil Science Society of America** (SSSA), a longtime program partner, provides lessons, activities, fun facts, sites of interest, and soil definitions for the novice soil scientist online. These resources were promoted by the October event. Earth Science Week's 2016 activity calendar featured a "Rain and Soil" classroom activity courtesy of SSSA.

StudentEnergy.org provided a card for the Earth Science Week 2016 Toolkit featuring information on the program's energy science education resources available online.

For the Earth Science Week 2016 Toolkit, **UNAVCO** provided a ruler featuring a colorful display of earthquakes, plate motions, and plate boundary strain rate, as well as a URL where more information can be found online.

The staff of Earth Science Week exhibited at the Science Expo of the fourth **USA Science & Engineering Festival** in Washington, D.C., in April 2016. AGI geoscientists were among the leading science and engineering organization representatives on hand to offer free events and presentations promoting awareness of science. More than 350,000 attendees came to celebrate science at the Expo.

The **U.S. Bureau of Land Management** (BLM), a continuing Earth Science Week partner, provided a flyer including a dinosaur activity sheet for the 2016 educator kit. BLM also was the subject of Earth Science Week promotions, including its Classroom Investigation Series online. The Earth Science Week 2016 activity calendar included a "Groundwater Movement" activity from BLM.

The **U.S. Department of Energy**'s Office of Energy Efficiency & Renewable Energy website offered classroom activities and materials for K-12 science instruction, as program participants learned. Additionally, educators were invited to explore DOE's websites for the National Renewable Energy Laboratory. Earth Science Week participants were made aware that AGI's Center for Geoscience & Society produced education materials, including videos in English and Spanish, education guides, a "quick start" guide to energy literacy, lesson connections, and guidance on aligning energy literacy lessons with the *Next Generation Science Standards*. Essential Principles and Fundamental Concepts for Energy Education resources were made available on the DOE website. Teams of program participants were urged to enter DOE's new Geothermal Design Challenge, which invites young people to design an infographic that illustrates how geothermal energy is clean, safe, reliable, and renewable.

Earth Science Week also promoted awareness of the **U.S. Environmental Protection Agency**'s collection of free resources to enhance middle school students' understanding of climate change impacts on the United States' wildlife and ecosystems. The online toolkit includes case studies, activities, and videos based on climate science, environmental education, and stewardship information.

Overlapping Earth Science Week 2016, National Wildlife Refuge Week also was held October 11-17. The event, celebrating the richness of the 550 units that make up America's National Wildlife Refuge System, was sponsored once again by the **U.S. Fish and Wildlife Service** (FWS), an Earth Science Week partner.

Earth Science Week participants learned about online education resources offered by the **U.S. Geological Survey** (USGS), a longtime Earth Science Week partner and supporter, as well as the thousands of free images and over 69,000 searchable publications such as maps,

Finalist entry to the Earth Science Week 2016 Photography Contest by Charles Carrigan.
Image credit: Charles Carrigan

books, and charts provided online by the agency. Also, USGS continued its leadership role as a founding partner of Geologic Map Day in 2016, providing support as well as its National Geologic Map Database's MapView, which offers a mosaic view of published geologic maps. Finally, program participants were invited to take part in the first Earth Observation Day, a STEM education event celebrating the Landsat mission of USGS and NASA.

Earth Science Week promoted awareness of **Windows on Earth**, an online educational project that features photographs taken by astronauts on the International Space Station. The site is operated by TERC, an educational non-profit, in collaboration with the Association of Space Explorers (the professional association of flown astronauts and cosmonauts), the Virtual High School, and CASIS (Center for Advancement of Science in Space). The images help show Earth from a global perspective.

PLAN!T NOW's **Young Meteorologist Program** takes students on a severe weather preparedness adventure while learning about severe weather science and safety. Promoted by Earth Science Week, the program was developed in partnership with the National Oceanic and Atmospheric Administration's National Weather Service and the National Education Association. Young Meteorologists are shown opportunities to put their knowledge to work in hands-on activities and community service projects.

2016 Earth Science Week poster.
Image credit: Poster design by Angela Terry for AGI

Earth Science Week Toolkits

Across the country, AGI distributed some **16,000 Earth Science Week Toolkits** to teachers and geoscientists in 2016. The number of AGI member societies requesting complimentary Earth Science Week Toolkits for distribution was 21 (up from 18), and the number of state geological surveys requesting complimentary kits for distribution was 30. As in past years, thousands of kits also were distributed through program partners including USGS, NASA, the National Park Service, and AAPG Student Chapters. Hundreds of kits were shipped free to geoscience department chairs at colleges and universities nationwide. Toolkits were shipped to program participants in all 50 states and a number of countries overseas.

The 2016 toolkit featured AGI's traditional Earth Science Week poster, education and outreach flyer, and school-year calendar showcasing geoscience classroom investigations and important dates of Earth science events. Program partners' contributions made the Earth Science Week 2016 Toolkit one of the richest in recent years.

Web Resources

The Earth Science Week website (www.earthsciweek.org) delivers essential resources for educators throughout the year. A recent redesign and new features made it easier than ever to find key programs, publications, and educational tools. The redesigned site offers a more attractive presentation, streamlined functionality, and user-friendly interface.

As in past years, the **Earth Science Week website** was updated regularly to reflect the new theme, contests, proclamations, events, initiatives, and classroom activities for 2016. The entire site received more than 675,000 page views. Within the site, Classroom Activities pages received over 475,000 views. Contests pages received more than 17,000 views. Plan an Event pages received over 4,700 views.

Released a few months prior to 2016, AGI's **Earth Science Week promotional video** trumpeted the importance of the geosciences and the celebration's role in promoting public awareness. This brief, exciting, eye-popping video answers key questions: Why is Earth science a big deal? How does Earth Science Week help promote learning and teaching about the subject? And what can students, educators, community partners, and others do to get involved?

Science teachers and students were invited to go online to use a new educational resource of the Earth Science Week website, the **Our Shared Geoheritage** page, which features educational material on geoscience heritage. Supporting the Earth Science Week 2016 theme of "Our Shared Geoheritage," this new page on the program website links users to recommended resources including downloadable reports, articles, blogs, geoheritage locations, and learning activities. The new page also features geoheritage-related classroom activities and links to information on geoheritage in every state.

Another recent addition is an Earth Science Week page helping students explore geoheritage through visualizations. The **Visualizing Earth Systems** page links users to dozens of recommended visualizations dealing with energy, climate, minerals, water, hazards, and other topics. In addition, the page links users to overviews of these topics provided by AGI's Critical Issues Program. Users are invited to help improve the page by sharing their favorite Earth science visualizations.

A new **Earth Science Week Event Registry** heightened visibility nationwide for events such as exhibits, tours, lectures, and open houses. To register an event, the host simply provides a few key details online. All registered events are listed on Earth Science Week's Events In Your Area page.

Another resource added to the website was the **Earth Science Week Innovations** page. As program participants learned, 2016 brought a new Our Shared Geoheritage resource, the One Shared Place video competition, Earth Observation Day, an expanded roster of Citywide Celebrations, and more. As the new Innovations page shows, Earth Science Week has added to its bursting clearinghouse of educational resources, activities, and initiatives available to participants year after year throughout its history.

Once again AGI offered four quarterly **Earth Science Week Webcasts** in 2016, expanding the program's use of online formats and media for public outreach. The free webcasts provided lively overviews of Focus Days (spring), Contests (summer), the Toolkit (autumn), and the Roy Award (winter). Each roughly five-minute tutorial includes a wealth of online links, which viewers can click during the narrated presentation to review available resources.

Program participants were encouraged to visit the continually updated Earth Science Week **Classroom Activities** page for more than 130 free learning activities, most of them contributed by leading geoscience agencies and groups. Activities are organized and searchable by various criteria, including specific Earth science topics. To find the perfect activity for a lesson, teachers can search by grade level and science education standard. Maybe most useful, they also can search among 24 categories of Earth science topics, from energy and environment to plate tectonics and weathering. This may be why Classroom Activities rank as one of the program's most popular online offerings, with 99 percent of survey respondents rating it as "useful" or "very useful."

AGI provided a set of free online videos and other electronic resources to help students, educators, and others explore the "big ideas" of Earth science during Earth Science Week 2016 and throughout the year. **Big Ideas Videos** bring to life the nine core geoscience concepts that everyone should know. The Earth Science Literacy Initiative, funded by the National Science Foundation, codified these principles. The videos are available on YouTube and TeacherTube. The Earth Science Week website also provides dozens of classroom activities linked to the "big ideas."

Those hosting events during Earth Science Week 2016 were invited to let people know about it at **Events in Your Area.** This web page provided information on events taking place through program partners in each state. In addition, events and groups could be listed in **Earth Science Organizations,** an online map that offers clickable links to Earth Science Week events taking place at parks, museums, science and technology centers, university geology departments, local geological societies, and other nearby locations.

A page dedicated to **Geoscience Career, Scholarship, and Internship Resources** was added to the program website. Another page of **Links** was included to provide external connections to sites featuring resources on key topics such as chemistry, climate, drought, earthquakes, energy, floods, hurricanes, landslides, sinkholes, soil, tornadoes, tsunamis, volcanoes, and wilderness fires.

Finally, Earth Science Week makes ample use of online social networking to reach new audiences, especially young people. The program's presence on **Facebook,** the Internet's most popular networking site, includes an Earth Science Week Fan Page. In addition, web surfers are invited to receive geoscience news, resources, and opportunities by following Earth Science Week on **Twitter.** Tweets are sent frequently, whenever there is valuable news or information to share. The number of people learning about Earth Science Week through social media soared in 2016, as tens of millions of people received program information from AGI and program partners through Facebook, Twitter, and Pinterest.

Newsletter

The monthly **Earth Science Week Update** newsletter reached over 5,300 teacher, student, and geoscientist subscribers in the past year. The electronic newsletter kept planners and participants up-to-date on Earth Science Week planning at the national level, encouraged participation in local areas, and provided news on geoscience topics of interest to participants.

Besides highlighting worthwhile resources, these monthly e-mail updates reinforce the belief that geoscience education is a priority throughout the year, not only during one week each October. It is little wonder that the e-newsletter remains a popular online offering, with 93 percent of survey respondents rating it as "useful" or "very useful."

Contests

Two of the four winning teams from the 2016 "One Shared Place" video contest. At left, "The Science Geeks" and at right, "The Science Superbs."

Image credits: both by teacher Karen Folleras, Academy International Elementary in Colorado Springs, CO

AGI held contests in connection with Earth Science Week for the 16th year. Contests were designed to encourage K-12 students, teachers, and the general public to become involved in the celebration by exploring artistic and academic applications of Earth science. Earth Science Week expanded eligibility for its photo contest to allow international members of **AGI Member Societies** and **AGI International Affiliates** to participate.

Three traditional contests continued to provide ways for many people to participate in Earth Science Week. Photos, art, and essays were submitted by hundreds of people. Each first-place winner received $300 and a copy of AGI's The Geoscience Handbook. Entries submitted by winners and finalists were posted online.

Sophia Findley of Kapolei, Hawaii, won first place in the visual arts contest with a creative and colorful drawing of Diamond Head Volcano, an example of geoheritage near where she lives. Finalists were David De Costa, Jenna Eutsey Shrish Karthikeyan, and Taylor Redmond. Students in grades K-5 made two-dimensional artworks illustrating the theme **"Seeing Earth Heritage."** Lily Wolfe of San Carlos, California, won first place in the photo contest with her image of seaside geology along with underwater life. Finalists were Charles Carrigan, Calder Laidlaw, Katy Sprinkle, and Vineela Vandanapu. Submissions illustrated the theme **"Our Heritage in Earth Systems."** Logan Flanagan of Virginia Beach, Virginia, won first place in the essay contest with his paper on "An Untouched World." Finalists were Lalitha Aiyar, Isabella Fernandez, Emma Hutchinson, and Brent Wang. Students in grades 6-9 wrote essays of up to 300 words addressing this year's theme, **"Sharing and Caring for Our Geoheritage."**

In addition, AGI invited teams of educators and students to enter its new **One Shared Place video contest.** Each team submitted a 30- to 90-second original video informing viewers about an outdoor place deemed special in terms of geoheritage and geoscience. By displaying entries for public viewing online, AGI aimed to enhance public understanding of these topics. The contest, presented by AGI in partnership with the National Park Service, was open to teams anywhere in the world. A contest overview was provided in a brief One Shared Place Introduction Video made available via social media and the competition website.

One Shared Place named four winning teams: **Deal Island,** consisting of educator Kolby Noble and students Ashlynn Merrit, Matthew Price, Lasadia Skelley, Dawson Daniels, Nathan Bradford, Jazmine White, Madison Mechalske, Ethan Scott, Alyssa Moisan, Kenny Stachawski, and Paige Barnes; **Rock Hard,** consisting of educator Margaret Salter and students Nick Urbanic, Hayes Densman, Savannah Urbanic, and Drake Densman; **Science Geeks,** consisting of educator Karen Folleras and students Audrey Chang, Emilia Lazarova, Gousikaa Suresh, and Ashlyn Laubach; and **Science Superbs,** consisting of educator Karen Folleras and students Prarthana Mathad, Peter Richardson, Ashton Holmes, Beckett Shanon, Aashay Deshpande, and Lucian Ensell. Winning videos were posted online.

Earth Science Teacher Award

For the ninth consecutive year, AGI and the AGI Foundation offered the **Edward C. Roy, Jr. Award for Excellence in K-8 Earth Science Teaching**. Dianna McDowell, a teacher at Kemps Landing Old Donation School in Virginia Beach, Virginia, was named the 2016 recipient of the prestigious award. McDowell, who earned her master's degree in curriculum and instruction from the University of Virginia, started as a Naval Oceanography Officer and switched careers to become a middle-school Earth science teacher in order to foster natural curiosity about the world.

McDowell received a $2,500 prize and an additional grant of $1,000 to enable her to attend the National Science Teachers Association 2016 National Conference to accept the award during a reception hosted by the National Earth Science Teachers Association. Finalists for the award were Ella Bowling of Mason

County Middle School in Maysville, Kentucky, and James Christopher Spiegl of Montgomery Bell Academy in Pegram, Tennessee.

The award recognizes one classroom teacher from kindergarten to eighth grade for leadership and innovation in Earth science education. This award is named in honor of Dr. Edward C. Roy, Jr., a past president of AGI and strong supporter of Earth science education. In addition to U.S. teachers, instructors throughout the United Kingdom were invited to compete for the prize. U.K. teachers were provided with detailed guidance on entering the competition by AGI and **The Geological Society of London**, a member society and Earth Science Week partner.

Events

Earth Science Week 2016 kicked off on Sunday, October 9, with the ninth annual **International EarthCache Day**. "EarthCaching" is a variation of a recreational activity known as geocaching, in which a geocache organizer posts latitude and longitude coordinates on the Internet to advertise a cache that geocachers locate using GPS devices. The activity has attracted over a million participants worldwide. When people visit an EarthCache, they learn something special about Earth science, the geology of the location, or how the Earth's resources and environment are managed there. EarthCaching has been developed by the Geological Society of America, a major program partner.

On Monday, October 10, educators and young people were encouraged to explore "big ideas" as part of **Earth Science Literacy Day.** The AGI "Big Ideas of Earth Science" videos provided on YouTube and TeacherTube outline the core concepts of geoscience, as codified by the Earth Science Literacy Initiative with support from the National Science Foundation. To help teachers and students use the videos, the Earth Science Week website offers dozens of related classroom activities.

One of the highlights of recent years' Earth Science Week celebrations has been **"No Child Left Inside" Day,** an event that in its inaugural year engaged some 500 students in outdoor learning activities and received coverage by news media from NBC to NPR. In 2016, students and educators nationwide were invited to take part on the Tuesday of Earth Science Week, October 11. AGI's online NCLI Day Guide provided everything needed to plan a local NCLI Day event. The free guide provided 17 outdoor activities, as well as detailed recommendations for creating partnerships, planning logistics, reaching out to the local media, and following up in the classroom.

In addition, the Tuesday of Earth Science Week 2016 marked the first time that participants were invited to take part in **Earth Observation Day.** Previously celebrated at other times of the year, this October 11 event aimed to engage students and teachers in remote sensing as an exciting and powerful educational tool. The event was a STEM educational outreach event of AmericaView and its partners. AmericaView is a nationwide partnership of remote sensing scientists who support the use of Landsat and other public domain remotely sensed satellite data through applied remote sensing research, K-12 and higher STEM education, workforce development, and technology transfer. Participants made use of lessons and activities by AmericaView and other organizations, as well as additional Earth Observation Day resources, online.

Earth Science Week featured the return of a popular event on Wednesday, October 12, 2016. In partnership with the National Park Service (NPS), AGI helped conduct **National Fossil Day,** including activities and resources designed to celebrate the scientific and educational value of fossils, paleontology, and the importance of preserving fossils for future generations. NPS offered a website full of educational resources and information designed specifically for students and teachers. On the site's NPS Fossil Park Highlights page, visitors could find lesson plans developed to reflect state standards, fossil trading cards, videos about pygmy mammoths, special brochures, a virtual museum exhibit on dinosaurs, and more. NPS also held a National Fossil Day Art Contest. Finally, AGI collaborated with NPS partners and other geoscience organizations to conduct the seventh annual National Fossil Day event in Washington, D.C. NPS and the Smithsonian Institution collaborated to host the National Fossil Day Celebration at the Smithsonian's National Museum of Natural History.

Program participants were invited to join the Earth Science Week team in encouraging everyone — including women, minorities, and people with a range of abilities — to explore geoscience careers on **Geoscience for Everyone Day**, Thursday, October 13. Educators welcomed geoscientists into the classroom to speak. Geoscientists visited schools and volunteered at science centers. Others organized scout events, led field trips, and held special "Take Your Child to Work Day" events.

The aim was to open a young person's eyes to the world of Earth science. Doing so, participants supported the efforts of AGI member societies such as the Association for Women Geoscientists and the National Association of Black Geoscientists in raising awareness of the remarkable opportunities available to all young people in the Earth sciences. The program website directed participants to "Visiting Geoscientists: An Outreach Guide for Geoscience Professionals," a handbook co-produced by AGI and the American Association of Petroleum Geologists' Youth Education Activities Committee.

The fifth annual **Geologic Map Day** held on Friday, October 14, 2016, promoted awareness of the study, uses, and importance of geologic mapping for education, science, business, and a variety of public policy concerns. The final event for the school week of Earth Science Week 2016 was hosted by the U.S. Geological Survey and the Association of American State Geologists in partnership with AGI, along with additional partners including the National Park Service, the Geological Society of America, NASA, and Esri. Students, teachers, and the wider public tapped into the various educational activities, print materials, online resources, and public outreach opportunities for active participation. The Earth Science Week 2016 Toolkit contained a Geologic Map Day poster that provided geologic maps, plus step-by-step instructions for a related classroom activity. Additional resources for learning about geologic maps were featured on the Geologic Map Day web page of the Earth Science Week site. Activities nationwide, many led by state geologic surveys, spurred learning in schools.

Earth Science Week 2016 reached its climax with **International Archaeology Day** on Saturday, October 15. The event was a celebration of archaeology and the thrill of discovery. Every October, archaeological programs and activities for people of all ages and interests are presented by the Archaeological Institute of America and archaeological organizations across the United States, Canada, and elsewhere. Programs included activities such as a family-friendly archaeology fair, a guided tour of a local archaeological site, a simulated dig, and a lecture or a classroom visit from an archaeologist. In every case, interactive, hands-on International Archaeology Day programs provided the chance for participants to indulge their inner "Indiana Jones."

Citywide Celebrations

City-specific celebrations served as major centers of public awareness activities during Earth Science Week 2016. **Houston, Denver, Washington, D.C., and Richmond, Virginia** extended and deepened the reach of the nationwide event. In all these cities, AGI donated hundreds of Earth Science Week Toolkits and collaborated with geoscience organizations and public schools to provide special events, educational materials, online resources, and activities in schools and other settings.

AGI also conducted an **Earth Science Week Richmond Stakeholder Reception** to gather, honor, and facilitate continuing partnerships among government, nonprofit, and private-sector organizations in the geosciences in this Virginia city. Speeches, presentations, and discussions explored issues relevant to the geoscience community. To support the celebrations, AGI also donated dozens of Earth Science Week Toolboxes to schools in Richmond. These collections contained study guides, posters, disks, and other resources for Earth science education.

AGI's **Citywide Celebrations website** provided educators, students, and families with links to additional educational resources as well as other offerings in participating cities. Program participants nationwide were encouraged to collaborate with local partners to launch their own Citywide Celebration.

AGI Promotions

Earth Science Week promoted awareness of numerous AGI programs and resources of interest to Earth science educators, students, and enthusiasts, including AGI's Center for Geoscience & Society, the AGI Geoscience Workforce program, the William L. Fisher Congressional Geoscience Fellowship, AGI's NGSS Education Summit and Webinars, the Pulse of Earth Science website, the Visiting Geoscientists guide, the One Shared Place competition, AGI's Critical Issues Program, the Earth Science Organizations website, the Faces of Earth DVD, AGI's EARTH magazine, AGI's District Visit Days, the Why Earth Science video, and AGI's and the National Park Service's jointly published *America's Geologic Heritage: An Invitation to Leadership* (a prestigious American Graphic Design Award was presented in 2016 by the magazine Graphic Design USA to Angela Terry Design for the design of this publication).

Congressional Recognition

In a display of bipartisanship, U.S. senators from across the country came together in support of AGI's 19th annual Earth Science Week celebration. The coalition, which consisted of members of Congress from Alaska to Florida, formally introduced a Senate resolution officially designating the week of October 9-15, 2016, as Earth Science Week.

Senator Lisa Murkowski (R-Alaska), the senior Senator from Alaska and the Chairman of the Senate Energy and Natural Resources Committee, introduced S. Res. 562 on September 15. It formally passed the Senate by unanimous consent on September 28. The resolution emphasizes how a strong foundation in the Earth sciences contributes to our ability to serve as responsible stewards of the Earth and its resources.

Senators Maria Cantwell (D-Washington), Shelley Moore Capito (R-West Virginia), Bill Cassidy (R-Louisiana), Mazie K. Hirono (D-Hawaii), Bill Nelson (D-Florida), Gary Peters (D-Michigan), Angus King (I-Maine), Edward Markey (D-Massachusetts), and Martin Heinrich (D-New Mexico) joined Senator Murkowski as cosponsors of the resolution.

State Proclamations

Governors issued proclamations recognizing Earth Science Week 2016 in **12 states**: Alaska, Delaware, Illinois, Louisiana, Missouri, Nevada, New Mexico, North Dakota, Oklahoma, Pennsylvania, South Dakota, and Tennessee. Seven of these states have issued "perpetual proclamations," ensuring recognition of Earth Science Week every year.

Publicity and Media Coverage

AGI enlisted the support of a wide range of media in promoting awareness of Earth Science Week, resulting in unprecedented reach for promotional activities in 2016 and helping to lay a foundation for more coverage in years to come.

News Media: Earth Science Week news and events were covered by major national news organizations such as *Business Wire, Earthzine, E&P Magazine, Market Wired, Physician's Money Digest, Planeta,* and *Ripple News.* Additionally, the event was covered by international news publications and organizations including *The Daily Mail* of the United Kingdom, *Canberra Times* of Australia, *Earth Observatory Blog* of Singapore, *HTS News* of Saint Lucia, *Liveguide* of Australia, *Loop News* of Jamaica, the official city website of London, Canada, *Saint Lucia Times* of Saint Lucia, and *The Sydney Morning Herald* of Australia.

Coverage of Earth Science Week programs and activities was provided by local news organizations such as *The Bemidji Pioneer* of Minnesota; *Chadds Ford Live* of Pennsylvania; *The Chattanoogan* of Tennessee; *Chester County Press* of Pennsylvania; *Cody Enterprise* of Wyoming; *Corpus Christi Caller Times* of Texas; *The Crimson White* of Alabama; *The Daily Sentinel* of Colorado; *Delaware Online* of Delaware; *Eugene Weekly* or Oregon; *Harker Heights Herald* of Texas; *Illinois RiverBender* of Illinois; *Inland Valley Daily Bulletin* of California; *Killeen Daily Herald* of Texas; *Ladue News* of Missouri; *Lancaster Online* of Pennsylvania; *Lawrence Journal-World* of Kansas; *Los Alamos Daily Post* of New Mexico; *Madison County Courier* of New York; *Martha's Vineyard Times* of Massachusetts; *Mid-Michigan Nature and Science* of Michigan; *Montana State* of Montana; *Nevada Today* of Nevada; *The News Journal* of Delaware; *North Texas e-News* of Texas; *Northern Minnesota Hometown Focus* of Minnesota; *Northern Nevada Business Weekly* of Nevada; *Paulding County Progress* of Ohio; *Rockford Register Star* of Illinois; *Sandusky Register* of Ohio; *Scottsbluff Star Herald* of Nebraska; *The Santa Fe New Mexican* of New Mexico; *Tennessee Today* of Tennessee; *UPMatters News* of Michigan; *Waco Tribune* of Texas; *Waterloo-Cedar Falls Courier* of Iowa; *Western Slope Now* of Colorado.

AGI distributed press releases to hundreds of newspapers, magazines, and other print media outlets. Many local and national publications ran items on Earth Science Week (for examples, see news clippings online). AGI staff also wrote articles for the Society of Exploration Geophysicists' *The Leading Edge* and the National Earth Science Teacher Association's *The Earth Scientist* magazine. Program Manager Geoff Camphire gave an interview for the American Association of Petroleum Geologists' *AAPG Explorer.* The articles highlighted Earth Science Week activities and theme.

More than 61,500 copies of the Earth Science Week 2016 poster, featuring a geoscience learning activity in addition to promotional content, were distributed as inserts in publications carrying articles about the event, such as *The Earth Scientist, GSA Today, AAPG Explorer,* and AGI's *Earth* magazine.

Internet: Earth Science Week activities were highlighted on hundreds of Internet pages besides the official event site. Information promoting events and publicizing the celebration was featured by program partners and promoters including Aerotech News; American Geosciences Institute; American Geophysical Union; Australian Government Geoscience Australia; The Bemidji Pioneer; Business Wire; Canberra Times; CBS News; Chadds Ford Live; The Chattanoogan; Chester County Press; Cody Enterprise; Corpus Christi Caller Times;

The Crimson White; The Daily Mail; The Daily Sentinel; Delaware Online; E&P Magazine; Earth Observatory of Singapore Blog; Eugene Weekly; Mississippi State University; University of Missouri; Wyoming State Geological Survey; Hands on the Land; Harker Heights Herald; HTS News Saint Lucia; Illinois RiverBender; Inland Valley Daily Bulletin; KCGS News; Killeen Daily Herald; KOOL 107.9 FM; KQKY Hits 106; KSWO ABC 7 News; KVAL CBS 13; KXAN NBC; Ladue News; Lancaster Online; Lawrence Journal-World; Liveguide; London, Canada, Official City Website; Loop News; Los Alamos Daily Post; Madison County Courier; Market Wired; Martha's Vineyard Times; Mid-Michigan Nature and Science; Montana State University; NASA; National Science Foundation; Nevada Bureau of Mines and Geology; Nevada Today; The News Journal; North Texas e-News; Northern Minnesota Hometown Focus; Northern Nevada Business Weekly; Paulding County Progress; Physician's Money Digest; Planeta; PR Web; Ripple News; RivetedLit; Rockford Register Star; Sandusky Register; The Santa Fe New Mexican; Scottsbluff Star Herald; Speaking of Geoscience GSA Guest Blog; Saint Lucia Times; Tennessee Today; The South Texan TAMU Kingsville; The Sydney Morning Herald; Through the Sandglass; University of Missouri; UPMatters News; Waco Tribune; Washington Department of National Resources; Water Education Foundation; Waterloo-Cedar Falls Courier; and Western Slope Now. Earth Science Week was mentioned on hundreds of websites, social networks, and blogs.

Television/Radio: Earth Science Week was covered by television stations across the country, including CBS News; HTS News Saint Lucia; KCGS TV News of Utah; KSWO ABC 7 News of Oklahoma; KVAL CBS 13 of Oregon; and KXAN NBC of Austin. In addition, the event was covered by radio stations such as KQKY Hits 106 of Nebraska and KOOL 107.9 FM of Colorado.

External Evaluation of Earth Science Week 2016: Key Findings

Following the event, AGI secured an independent contractor, PS International, to complete a formal external evaluation of Earth Science Week 2016, as it has in past years. Some 8,000 program participants were invited to participate in a survey in the closing months of 2016, with a valid response rate of 2.3 percent.

Results were overwhelmingly positive. Comparing participation last year and plans for the next year, 86 percent of survey respondents said they anticipate either increasing or maintaining level participation. Respondents voiced enthusiasm despite noting challenges and barriers, including budget cutbacks, shrinking staffs, and competing priorities.

A large majority of participants (82 percent) once again said Earth Science Week offers opportunities for teaching and promoting Earth science that they wouldn't have otherwise. Participants commented on the "new information" and "teacher Kits."

Similarly, 85 percent said program resources and activities are "very" or "somewhat" important to educating students and others about Earth science. "The ESW resources in the teaching toolkits are wonderful resources," remarked one participant.

Most respondents find Earth Science Week and related resources highly useful, with 78 percent rating the program's overall usefulness as "excellent" or "good." When respondents were asked to rate 15 key items from the Earth Science Week 2016 Toolkit, all were rated "very useful" or "useful" by strong majorities of participants. Online offerings received similarly high marks.

Top-rated items were AGI classroom activities, ESW e-newsletter, Big Ideas videos, ESW Toolkit ordering information, Focus Days details, NASA visualization DVD and booklet on Earth science, NPS posters on geologic and air resources, 12-month school-year activity calendar, Geologic Map Day poser with geologic mapping activity, climate worksheet from American Meteorological Association, UNAVCO resource on Earth's gravity and rotation, ecology DVD and more from Howard Hughes Medical Institute, and material on energy science from Student Energy.

Participants said they were active during Earth Science Week. Many reported specific activities that were highly active. For example, 93 percent reported activities categorized as "most active" (e.g., field trips and outside lessons), "active" (e.g., external speakers and open house discussions), or "somewhat active" (e.g., lesson plans and kit distribution).

Asked how Earth Science Week might be improved, the largest share of respondents (36 percent) advocated additional program materials and activities. Additional suggestions included increasing communication and promotion. AGI uses these evaluation findings to fuel continual improvement of the program.

Earth Science Week Sponsors

United States Geological Survey

National Aeronautics and Space Administration

National Park Service

American Association of Petroleum Geologists Foundation

American Geophysical Union

Society for Mining, Metallurgy, and Exploration

Howard Hughes Medical Institute

Association of American State Geologists

Geological Society of America

Archaeological Association of America

AmericaView

ExxonMobil

Esri

Earth Science Week Program Partners

American Association of Petroleum Geologists Foundation

American Geophysical Union

American Geosciences Institute

American Institute of Professional Geologists

AmericaView

Archaeological Institute of America

Association for Women Geoscientists

Association of American State Geologists

Critical Zones Observatories

Esri

ExxonMobil

Geological Society of America

Google

Howard Hughes Medical Institute

Incorporated Research Institutions for Seismology

Minerals Education Coalition

National Earth Science Teachers Association

National Energy Education Development Project

National Science Teachers Association

Paleontological Research Institution

Paleontological Society

SEED: Schlumberger Excellence in Educational Development

Society for Mining, Metallurgy, and Exploration

Society of Exploration Geophysicists

Society of Petroleum Engineers

Soil Science Society of America

Student Energy

UNAVCO

U.S. Bureau of Land Management

U.S. Department of Energy

U.S. Fish and Wildlife Service

U.S. Geological Survey

U.S. National Aeronautics and Space Administration

U.S. National Oceanographic and Atmospheric Administration

U.S. National Park Service

U.S. National Weather Service

Earth Science Week 2016 Events and Activities by State and Territory

Alabama

- The Alabama Geological Survey distributed 50 complimentary Earth Science Week Toolkits to science educators for use in the classroom.
- The University of Alabama's Smith Hall Museum of Natural History hosted a National Fossil Day event celebrating fossil appreciation and stewardship. The public was invited to meet a paleontologist and view fossils.

Alaska

- The governor of Alaska issued a perpetual proclamation of Earth Science Week.
- The Alaska Division of Geological & Geophysical Surveys distributed Earth Science Week Toolkits to science educators.

Arizona

- Grand Canyon National Park hosted its annual Earth Science Week and National Fossil Day celebration. Activities included ranger-led hikes exploring geologic time and paleontology lectures.
- The Pueblo Grande Museum in Phoenix hosted an International Archeology Day event in collaboration with the Central Arizona Society of the Archeological Institute of America that featured archeological demonstrations, children's activities, tours, and more.

Arkansas

- The Arkansas Geological Survey promoted Earth Science Week among teachers and citizens with the distribution of related instructional materials.

California

- The California Geological Survey distributed 50 Earth Science Week Toolkits to educators at the National Geoscience Teachers Association's fall conference.
- KQED recommended five "geological outings" that Californians could take to enjoy the natural world and learn about Earth science across the state during Earth Science Week.
- Through NASA's education outreach, California teachers were shown videos on how NASA's Environmental Control and Life Support System works to provide the International Space Station with water and air. Teachers at a workshop learned about research projects such as CalWater and Operation IceBridge, which are occurring at NASA Armstrong Flight Research Center.
- Lily Wolfe of San Carlos, California, won first place in the Earth Science Week photo contest with her image of seaside geology along with underwater life.
- More than 300 people attended the International Archaeology Day and California Archaeology Month Expo held in Riverside. The archaeology fair included 28 exhibitors from academic, government, and private organizations. Attendees learned about mapping using a total station, making stone tools, analyzing artifacts, and more.

Colorado

- The Geological Society of America encouraged geocachers around the world to participate in International EarthCache Day during Earth Science Week.
- Museums of the West held events featuring free educational activities for families across Colorado.
- Teams of educators and students entered Earth Science Week's new One Shared Place video contest. Two of four winning teams were Science Geeks and Science Superbs, led by educator Karen Folleras of Colorado Springs.
- Many events were held across Colorado throughout Earth Science Week, including geologist-lead hikes in Fruita, and Golden, as well as a number of events sponsored by Museums of Western Colorado, Colorado Canyons Association, The John McConnel Math & Science Center, and more.
- Earth Science Week 2016 Toolkits were distributed to students, teachers, and others by representatives of the American Institute of Professional Geologists, the Geological Society of America, the National Association of Earth Science Teachers, and the Society of Mineral Museum Professionals.

Connecticut

- The Connecticut Geological Survey promoted Earth Science Week and distributed kits among educators in the state.

Delaware

- The governor of Delaware issued a perpetual proclamation of Earth Science Week.
- The Delaware Geological Survey promoted Earth Science Week by distributing informational materials.
- The Delaware Geological Survey recommended fossil hunters visit the Chesapeake and Delaware Canal to search for the state fossil, a belemnite, to celebrate National Fossil Day.

District of Columbia

- The National Park Service and AGI hosted exhibits at the National Fossil Day celebration at the Smithsonian Nation Museum of Natural History.
- Numerous federal agencies, including the U.S. Geological Survey, NASA, and the National Park Service, supported Earth Science Week with the provision of teaching materials, distribution of kits, and events and activities.
- Earth Science Week 2016 Toolkits were distributed to students, teachers, and others by representatives of the American Geophysical Union.

Florida

- The Florida Geological Survey promoted Earth Science Week by distributing educational materials.
- Teams of educators and students entered Earth Science Week's new One Shared Place video contest. One of four winning teams was Rock Hard, led by educator Margaret Salter of Milton.
- Kaikea Nakachi, a senior in biological oceanography at Florida Tech, talked on YouTube about his passion for sharks, how growing up in Hawaii helped shape his studies, and how he plans on using his research to help others.

Georgia

- Earth Science Week 2016 Toolkits were distributed to students, teachers, and others by representatives of AASP-The Palynological Society and the National Association of State Boards of Geology.
- Students at Wilkes Primary School in Washington held a No Child Left Inside Day event sponsored by the Iris Garden Club, featuring outdoor learning activities including soil classifications, tree ring observations, and leaf rubbings guided by soil scientists, foresters, geologists, and others.

Hawaii

- Sophia Findley of Kapolei, Hawaii, won first place in the Earth Science Week visual arts contest with a creative and colorful drawing of Diamond Head Volcano, an example of geoheritage near where she lives.

Idaho

- The Idaho Geological Survey distributed Earth Science Week educational materials.

Illinois

- The governor of Illinois issued a perpetual proclamation of Earth Science Week.

Indiana

- The Indiana Geological Survey promoted Earth Science Week among citizens across the state.
- Earth Science Week 2016 Toolkits were distributed to students, teachers, and others by representatives of the North American Commission on Stratigraphic Nomenclature.

Iowa

- The University of Northern Iowa and BMC Aggregates partnered for the "Sunday at the Quarry" event in Waterloo to offer the public opportunities to dig for fossils, search for geodes, and interact with Earth systems.

Kansas

- The Kansas Geological Survey obtained and distributed Earth Science Week Toolkits for educators.
- The University of Kansas Natural History Museum celebrated National Fossil Day with a number of activities for the public including fossil casting and an excavation of fossilized shark teeth. The public was also invited to learn about the ongoing excavation of a *Tyrannosaurus rex* skeleton.

Kentucky

- The Kentucky Geological Survey hosted its annual Open House to celebrate Earth Science Week. Attendees were given free kits and were invited to browse the rock and fossil collections.
- The Kentucky Geological Survey held its annual open house during Earth Science Week 2016. Visitors had opportunities to view rocks under microscopes, observe a demonstration of a scanning electron microscope, and more.

Louisiana

- The governor of Louisiana declared October 9-15 to be Earth Science Week.
- The Louisiana Geological Survey distributed Earth Science Week Toolkits to schools across the state.
- Poverty Point World Heritage Site in Pioneer hosted activities for International Archeology Day, including a guided tram tour with the station archaeologist, a demonstration of prehistoric tools, and more.

Maine

- Finalists in the Earth Science Week essay contest included Emma Hutchinson of Maine.

Maryland

- Teams of educators and students entered Earth Science Week's new One Shared Place video contest. One of four winning teams was Deal Island, led by educator Kolby Noble of Deal Island.

Massachusetts

- The Oak Bluffs Library held a National Fossil Day event for Earth Science Week. People were invited to and share their fossils with collectors and listen to presentations.

Michigan

- The Michigan Geological Survey promoted Earth Science Week and distributed kits among educators statewide.
- Michigan Technological University hosted an Energy Day event in September as part of its CareerFEST. Students had opportunities to learn about the energy industry and different job and career opportunities.

Minnesota

- The Minnesota Geological Survey promoted Earth Science Week by distributing kits to teachers in the state.
- Sudharshan Seneviratne, recipient of the 2013 Archaeological Institute of America's Award for Best Practice in Conservation and Heritage Management, spoke about "Cultural Heritage Preservation in Sri Lanka" at Macalester College in St. Paul.

- In celebration of International Archaeology Day, the Minnesota chapter of the Archaeological Institute of America welcomed graduate and undergraduate students who presented posters about their recent archaeological fieldwork or research projects related to archaeology and cultural heritage preservation.
- Earth Science Week 2016 Toolkits were distributed to students, teachers, and others by representatives of the National Association of Geoscience Teachers.

Mississippi

- The Mississippi Department of Environmental Quality promoted Earth Science Week and distributed kits among educators in the state.
- The Department of Geosciences at Mississippi State University celebrated Earth Science Week 2016 with an open house, which featured tours of the department and the Dunn-Seiller Museum's "Fossil Extravaganza."

Missouri

- The governor of Missouri declared October 9-15 to be Earth Science Week.
- The Missouri Geological Survey promoted Earth Science Week among teachers and citizens with the distribution of educational materials.
- An International Archaeology Day event co-sponsored by the St. Louis County Parks Department and the Missouri Atlatl Association invited the public to Fort Belle Fountaine County Park to listen to a discussion on prehistoric hunting and learn to use atlatl to throw spears.
- The University of Missouri celebrated International Archaeology Day, disseminating information on archaeology and related activities statewide.

Montana

- Montana State University's Museum of the Rockies invited children to earn "Junior Paleontologist" badges by participating in National Fossil Day activities.

Nebraska

- Agate Fossil Beds National Monument celebrated National Fossil Day with special events.
- The radio station KQKY Hits 106 of Nebraska promoted awareness of Earth Science Week and the availability of program educational materials.

Nevada

- The governor of Nevada issued a perpetual proclamation of Earth Science Week.
- The Nevada Bureau of Mines and Geology sponsored an Earth Science Week field trip, "A River Runs Through It - Geology Along the Truckee River from Reno to Pyramid Lake," an educational tour for all ages. On the Saturday of Earth Science Week, the bureau hosted its annual Open House at its Great Basin Science Library. The free event featured activities for kids including gold panning and identifying faults from aerial photographs.
- Tule Springs Fossil Beds National Monument celebrated National Fossil Day at the Las Vegas Museum of Natural History. The public had opportunities to observe paleontologists doing fossil prep, explore hands-on fossil carts, and witness a shark feeding demonstration.
- Girl Scouts of the Sierra Nevada took part in geoscience field trips led by Daphne LaPointe, Geologist Emerita at Nevada Bureau of Mines & Geology.

New Hampshire

- The New Hampshire Geological Survey promoted awareness of Earth Science Week among residents and community members.

New Jersey

- The New Jersey Geological and Water Survey distributed Earth Science Week Toolkits and Geologic Map Day materials to teachers.

New Mexico

- The New Mexico Bureau of Geology distributed Earth Science Week kits among teachers statewide.
- The governor of New Mexico declared October 9-15 Earth Science Week.
- The New Mexico Museum of Natural History & Science celebrated National Fossil Day with a special program, "Afternoon Science Adventures," featuring hands-on fossil activities, a fossil identification station, and a presentation where museum staff and volunteers displayed their ongoing work on a recently discovered baby Pentaceratops dinosaur fossil.
- The Farmington Convention and Visitors Bureau encouraged residents and tourists to visit nearby Four Corners sites for International Archaeology Day. Sites included the Salmon Ruins, the Bisti/De-Na-Zin Wilderness area, the Simon Canyon Ruin, and the Tapacito Ruin.
- Earth Science Week 2016 Toolkits were distributed to students, teachers, and others by representatives of the National Speleological Society.

New York

- Earth Science Week 2016 Toolkits were distributed to students, teachers, and others by representatives of the Paleontological Research Institution.

North Carolina

- The North Carolina Geological Survey promoted awareness of Earth Science Week among residents and community members.
- Earth Science Week 2016 Toolkits were distributed to students, teachers, and others by representatives of the History of Earth Sciences Society.

North Dakota

- North Dakota's governor issued a perpetual proclamation of Earth Science Week.

Ohio

- Earth Science Week 2016 posters were included in the fall issue of "The Ohio Earth Scientist" by OESTA, which were distributed to Ohio geoscience teachers.
- The Ohio Department of Natural Resources (ODNR) Division of Geological Survey distributed Earth Science Week Toolkits to schools and others.
- The Ohio Geological Survey celebrated Earth Science Week by leading the public on geologic hikes meant to explore the effects of the Ice Age on the landscape.
- Earth Science Week 2016 Toolkits were distributed to students, teachers, and others by representatives of the Association of Environmental & Engineering Geologists and the National Ground Water Association.
- In celebration of National Fossil Day, Ohio State University and the Ohio Geological Survey ran a geologic tour of the fossil-rich building stones of Capitol Square in Columbus.
- The Cleveland Museum of Natural History hosted Ohio Geologic Survey geologists Mark Jones and Julie Bloxson, who discussed archeology, geology, mineralogy, and more.

Oklahoma

- Oklahoma's governor issued a perpetual proclamation of Earth Science Week.
- The Oklahoma Geological Survey distributed Earth Science Week Toolkits to educators and students.
- The Sam Noble Museum in Norman celebrated International Archaeology Day with family-oriented activities including flint knapping and atlatl demonstrations.
- Earth Science Week 2016 Toolkits were distributed to students, teachers, and others by representatives of the Society for Sedimentary Geology.

Oregon

- The Oregon Department of Geology and Mineral Industries released an interactive geologic map of Oregon for Geologic Map Day. The application, a decade in the making, allows the public the geologic story of Oregon and access supplementary information with a click of a mouse.

Pennsylvania

- The Pennsylvania Earth Science Teacher Association promoted awareness of National Fossil Day among science educators in Pennsylvania.
- Earth Science Week Toolkits were distributed among science teachers statewide by the Pennsylvania Geological Survey.
- The governor of Pennsylvania declared October 9-15 to be Earth Science Week.
- The Academy of Natural Sciences of Drexel University in Philadelphia hosted an event for National Fossil Day that featured talks from experts on fossil formation, interactive fossil displays, and more.
- Benjamin Edwards of Dickinson College in Carlisle spoke on Iceland's Eyjafjallajökull volcano and other volcanoes hidden under glaciers on behalf of the National Science Foundation.
- Earth Science Week 2016 Toolkits were distributed to students, teachers, and others by representatives of the Association of Earth Science Editors.
- The Newlin Grist Mill invited the public to participate in an archaeology dig to celebrate International Archeology Day.
- The Lancaster County Park's Environmental Center hosted a hike to investigate the geology of Chickies Rock and to discuss the natural history of the Susquehanna River.

Puerto Rico

- Earth Science Week Toolkits were distributed to science educators and outreach contacts in Puerto Rico.

Rhode Island

- The Roger Williams Park Museum of Natural History and Planetarium held its "Fossil Frenzy Weekend" to celebrate National Fossil Day. Visitors viewed fossils from the museum's vaults, went on a fossil quest, did a fossil puzzle, and made their own trilobite masks and more.

South Carolina

- The South Carolina Geological Survey celebrated Earth Science Week and Geologic Map Day by distributing Earth Science Week Toolkits to educators.

South Dakota

- South Dakota's governor issued a perpetual proclamation of Earth Science Week.

Tennessee

- The governor of Tennessee declared October 9-15 to be Earth Science Week.
- The Tennessee Division of Geology distributed Earth Science Week Toolkits to educators.
- The University of Tennessee's McClung Museum with support from the Archeological Institute of America hosted the "Can You Dig It?" event to celebrate International Archeology Day and National Fossil Day.

Texas

- The Austin Earth Science Week Consortium held a career event for middle school students. Earth science professionals gave presentations about their careers, and students were able to participate in hands-on activities through various exhibits. Students learned about petroleum, solar energy, industrial mining resources, surface and ground water resources, weather and climate, archeology, astronomy, and environmental science.
- The Houston Geocaching Society celebrated International Earth Cache Day.
- The Houston Geological Society held various events for Earth Science Week with partners in the Houston area, including field trips and a special celebration at the Houston Museum of Natural Science.
- The Energy Day Festival, sponsored by more than 100 businesses in Houston, brought thousands of people out to view demonstrations, learn about energy science, and play games to compete for prizes and giveaways.
- The Just Energy Foundation awarded top-performing students of the Science and Engineering Fair of Houston during the Energy Day Festival.
- The Houston Museum of Natural Science hosted Energy Day in Houston. The event offered hands-on activities for young people.
- Waco Mammoth National Monument invited the public to celebrate Geologic Map Day by completing a mapping activity.
- Texas Memorial Museum celebrated National Fossil Day in Austin, where the public was invited to learn about the sabre-toothed cats of Texas and other fossil finds.
- The Geology Department at Texas A&M University Kingsville kicked off a week of activities with an open house featuring a minerals display, a documentary on volcanoes, and an opportunity to view planets from the university's telescope observatory. At Hill Hall Observatory, attendees had opportunities to learn about astronomy by peering through telescopes at Venus, Mars, Saturn, and the moon.
- Cavazos Elementary School in Nolanville celebrated No Child Left Inside Day with outdoor science activities and demonstrations.
- Earth Science Week 2016 Toolkits were distributed to students, teachers, and others by representatives of the Petroleum History Institute.
- The National Parks Service recognized the City of Ladonia in Fannin County, Texas as a sponsoring participant of National Fossil Day. The Dallas Paleontological Society, and the Ladonia Fossil Park held additional Earth Science Week activities.
- Students at Cavazos Elementary School in Nolanville participated in No Child Left Inside Day by taking part in a number of engaging science activities around their school grounds.
- Entries to the National Fossil Day Art Contest from Reicher Catholic High School in Waco won first place, second place, and honorable mention.

Utah

- Dinosaur National Park celebrated National Fossil Day with events throughout Earth Science Week.
- The Utah Geological Society distributed Earth Science Week Toolkits to educators and students.

Vermont

- The Vermont Geological Survey shared geoscience materials to celebrate Earth Science Week.

Virginia

- The Virginia Department of Mines, Minerals and Energy distributed Earth Science Week Toolkits to teachers in the state.
- Logan Flanagan of Virginia Beach, Virginia, won first place in the Earth Science Week essay contest with his paper on "An Untouched World."
- Earth Science Week 2016 Toolkits were distributed to students, teachers, and others by representatives of the American Institute of Hydrology and the Society for Organic Petrology.

Washington

- Earth Science Week Toolkits were distributed among science teachers statewide by the Washington State Department of Natural Resources.
- Citizens were invited to celebrate all that comes from that ground at the annual Rock Auction in the Washington State Geology Library in Olympia. Funds raised from the "Rock-tion" benefitted the Friends of the Washington Geology Library.
- Nearly 600 people came out to attend the Archaeological Institute of America's Spokane Local Society day-long event, "Rome and its Legions," featuring a troupe of Roman reenactors demonstrating Roman culture and hosted at Gonzaga University.

- Pacific Lutheran University's Anthropology Club drew visitors for a celebration of International Archaeology Day. The club created informational poster boards, games and networking opportunities for those interested in archaeology.

West Virginia

- Senator Shelley Moore Capito of West Virginia joined Senator Lisa Murkowski and fellow legislators in a bipartisan show of support for the geosciences. The coalition introduced a Senate resolution officially designating the week of October 9-15 as Earth Science Week. It formally passed the Senate by unanimous consent in September. The resolution emphasizes how a strong foundation in the Earth sciences contributes to our ability to serve as responsible stewards of the Earth and its resources.

Wisconsin

- Earth Science Week 2016 Toolkits were distributed to students, teachers, and others by representatives of the Soil Science Society of America.

Wyoming

- The Wyoming State Geological Survey disseminated Earth Science Week 2016 Toolkits among students and educators statewide.
- The Wyoming State Geological Survey and University of Wyoming Geological Museum celebrated with a geoscience event on October 15.
- The Buffalo Bill Center of the West in Cody held a field trip for local elementary school students on International Archeology Day. Students learned differences between paleontology and archeology.

International Events

Australia

- Geoscience Australia's Earth Science Week 2016 celebrated the theme of "Our Shared Geoheritage."
- Australians participated by touring the National Fossil and Mineral Collection, celebrating National Fossil Day, and helped students get involved with Antarctic geoscience education activities.
- As part of its Earth Science Week celebrations, Australia added a 10-tonne Moruya granite boulder to its National Rock Garden in the capital, Canberra.
- The Geological Survey of New South Wales hosted an Open Day in honor of Earth Science Week.

Brazil

- Eighty students visited the Geodiversity Museum at the Geoscience Institute of the Federal University of Rio de Janeiro.

Cameroon

- The Geo-Environmental Research Association hosted the third Earth Science Week celebration in Cameroon. Geology and geoscience education was provided to students.

Canada

- The Oshawa Museum in Ontario partnered with Trent University Durham and Scugog Shores Museums to host an International Archeology Day event featuring interactive displays, engaging activities, and lectures on archeology.
- The Royal Ontario Museum celebrated International Archaeology Day with a free lecture by antiquities scholar Erin Thompson, discussing the effects of the art market on antiquities. The event included other activities and exhibits also.

Colombia

- The Archeological Institute of America was joined by over 100 Collaborating Organizations to celebrate International Archeology Day in cities all over the world including Bogotá.
- Events were held at the University of Pamplona in Cucuta. Participants included students and university staff.
- The National University of Colombia at Medellin also held activities with four schools in the Medellin metropolitan area.
- The Colombian Geological Society worked with the Ministry of Education, Ministry of Environment, and Institute of Hydrology, Meteorology and Environmental Studies on an initiative related to climate change and Earth Science Week.

Czech Republic

- The City of Prague Museum celebrated International Archeology Day by hosting a Karolinum event to promote awareness of archeologists' work.

Guyana

- The University of Guyana celebrated archaeology and the thrill of discovery by underscoring high points in archaeology, including not only Guyanese archaeology but also aspects of Mayan culture. Participants learned to make hot chocolate like the Mayans and about tools used by archaeologists.

Ireland

- Earth Science Week events took place in museums, libraries, schools, universities, geoparks and more, including urban geowalks, family activity days, talks, conservation days and fieldtrips.

Jamaica

- Loop News promoted public awareness of geologic hazards as part of Earth Science Week.

Mexico

- AAPG Chapter representatives worked with administrators to conduct activities at National Autonomous University of Mexico (UNAM) high schools and support the UNAM Geology Institute's TerraMovil (Mobile Earth) initiative, which puts Earth science education in primary and secondary schools.

Peru

- The AAPG Young Professionals Chapter and Peruvian Geological Society hosted Earth Science Week conferences with live streaming via Facebook. Approximately 20 people attended the talks held in the Geological Society auditorium, and more than 500 connected to the live streaming.
- International Archeology Day was celebrated in Nivín, in the Casma Province, Ancash Region. Bringing together students, teachers and parents from this rural community, the activity was carried out at the archaeological site of Nivín. This celebration promoted respect for the culture and care of natural resources of Nivín. Museum exhibits were open to the visiting public.

Russia

- Along with more than 100 Collaborating Organizations, the Archeological Institute of America promoted celebrations of International Archeology Day in cities around the globe including Samara, Russia.

Saint Lucia

- The Caribbean nation of Saint Lucia celebrated Earth Science Week for the first time with activities aimed at promoting awareness of geologic hazards. Participating organizations included the West Indies Seismic Research Centre, the National Emergency Organization of Saint Lucia, and the Soufriere Regional Development Foundation.

Singapore

- As part of its Earth Science Week activities, the Earth Observatory of Singapore blog featured accounts of two of its scientists' most memorable experiences in the field.

Trinidad & Tobago

- Under a collaborative effort by the University of the West Indies Seismic Research Centre and other area organizations, experts offered workshops and seminars on geoscience in celebration of Earth Science Week.
- AAPG Young Professionals organized a geology hike for students.

United Arab Emirates

- Through Earth Science Week 2016, program participants learned about and were encouraged to take part in SciFest Dubai, a festival to promote creativity, critical thinking, innovation, and STEAM education, including geoscience education. In addition to organizing the annual Dubai Science Festival, SciFest Dubai also promoted scientific literacy through workshops, think tanks, projects, conferences, community initiatives, innovative curricula, and radio shows.

United Kingdom

- The Geological Society of London (GSL) supported a number of Earth Science Week events including geowalks and a school workshop.
- UK Earth Science Week and GSL sponsored a geologic tour of Anglesey, an island in Wales.
- GSL also launched a photography competition for Earth Science Week with the theme "Earth Science in Action." Winning photos are included in GSL's 2017 calendar.
- More than a thousand visitors at the Thomas Hardye School in Dorchester, England, enjoyed the Family Science Festival, hosted by the Dorchester Area School's Partnership. Twenty laboratories offered demonstrations and attractions designed to introduce youngsters to science in celebration of National Biology Week and Earth Science Week.

Uruguay

- Earth Science Week 2016 Toolkits were distributed to students, teachers, and others by Professor Nelly Mañay on behalf of the International Medical Geology Association.

Made in the USA
Monee, IL
05 November 2026

67377161R00017